Introducing Enamelling

Introducing Enamelling
Valerie Conway

B. T. Batsford Limited *London*
Watson-Guptill Publications *New York*

© Valerie Conway 1970

First published 1970

7134 2427 3
Library of Congress Catalog Card Number 77–94230

Printed in Denmark by
F. E. Bording Limited, Copenhagen
for the Publishers
B. T. Batsford Limited
4 Fitzhardinge Street, Portman Square, London W1 and
Watson-Guptill Publications
165 West 46th Street, New York, NY 10036

Contents

Introduction	7
Enamelling as a craft in schools	8
Basic equipment	9
Preparation of enamels	13
Preparation of adhesives	15
Pendants and name tags	16
Masking solution	
Copper protector	22
Swirling	23
Leaf stencil pendant	25
Counter enamelling	28
Cloisonné	31
A *cloisonné* pendant using silver *cloisons*	34
Enamelled copper and silver brooch	37
Enamel and copper wire brooch	38
Pliqué à jour pendant and earrings	40
Metal foils	42
Wet application of enamel	45
Lustres	46
Making a copper dish	47
The use of firescale for decoration	62
Transfer designs	
Printing	64
Silk screen printing	66
Group projects for schools	68
1 An enamelled steel panel	68
2 Copper and glass panel	68
Enamelled mobile	71
Sculpture and three-dimensional panels	73
Colour and design	76
Suppliers	86
Bibliography	88

Introduction

Enamel is a glasslike compound fused on to a metal base. There is little difference between glass, ceramic glaze and enamel, apart from the fact that enamel is made to withstand the stresses set up by the expansion and contraction of the metal during and after firing in the furnace. Without this, the enamel would crack off the metal during cooling.

Enamelling was practised in very early times. Greek sculpture of the fifth century B.C. showed areas of gold, inlaid with enamel, such as the gold drapery of the statue of Zeus by Phideas, which according to ancient history stood in the Temple of Olympia, and was enamelled with figures and flowers. Bronze heads have been excavated set with enamelled eyes. In the Far East the use of enamelling was widespread. This was in the form of *cloisonné*, the *cloisons*, or cells, being filled with a paste of enamel. In Great Britain the various enamelling techniques can be studied at the British and Victoria and Albert Museums: Celtic enamels in the Treasure of the Sutton Hoo at the British Museum; Byzantine and Champlevé enamels at the Victoria and Albert. Champlevé was practised during the Middle Ages and is essentially Gothic in style, the metal being engraved and the enamel placed in the indentations.

The type of enamelling produced today has evolved from Limoges. This originated in Limoges, France, during the end of the fifteenth century when enamellists of this period discovered a way of painting finely ground enamel on to an enamelled base. The majority of contemporary enamellists produce this type of enamel, that is to say, working directly on to the surface of the metal either by painting or sifting.

Enamelling as a craft in schools

In many schools, especially where there is a metalwork department, enamelling may be introduced as an additional craft. If the school already has a pottery section the kiln may be used for firing panels and larger pieces. However this should be a front loading kiln and as the elements are exposed it is essential to always switch off the current whilst introducing the piece into the kiln. If purchasing a kiln or enamelling furnace it is advisable to obtain one with a pyrometer attached as this shows the exact temperature of the interior of the kiln and is necessary for controlling firing results. In the metalwork department the braising torch can also be used for enamelling.

Alternatively a propane torch and a tin can will make a small improvised furnace. Sheet metal can be cut into strips with a metal guillotine or shears, to form pendants, brooches, and name tags. Cut into rectangles and squares they can be enamelled and reassembled to form a mural or panel as a joint project and shaped pieces can be wired to form mobiles. Copper tubing may be sawn to make rings or bracelets or silver soldered together, to form a three-dimensional panel. Small sculptures can be made from annealed copper sheet, enamelled and fired in the furnace or kiln.

As copper is expensive, mild steel can often take its place (see page 68). However if working on steel use opaque enamels only for the ground coat as transparent enamels can only be applied on pure metal, ie, copper, silver, gold and gilding metal (5 parts copper to 1 part zinc).

Basic equipment

One can start enamelling with very simple equipment and a few colours. A modest outlay would be sufficient to buy a small furnace, enamels and basic equipment. There are various firms which now cater especially for schools and the home craftsman and it is possible to obtain a basic kit, including a hot plate type of kiln, powdered enamels and copper blanks. (See page 86.)

A suggestion for basic colours is as follows,

OPAQUE (455 g, 1 lb, of each) black, white, scarlet, blue or turquoise, lime green and grey.
TRANSPARENT (455 g, 1 lb, of each) flux, gold, mauve, brown, turquoise and royal or deep blue.
LUMP ENAMELS (113 g, ¼ lb, of each) black, white, scarlet, turquoise (opaque); orange or gold, green, blue, mauve (transparent).

1a Muffle type enamelling furnace   1b Small enamelling furnace with pyrometer attached

2 Materials required for enamelling

1 Enamels   2 Lump enamel   3 Gum tragacanth
4 Roll of paper   5 Copper blanks   6 Brushes
7 Spatula and sgraffito tool   8 Sieve   9 Screw punch
10 Punch   11 Brass tongs   12 Round-nose pliers
13 Cap sieve   14 Carborundum stone   15 Steel stilts
16 Nichrome wire trivet

3 Larger front loading furnace

5 Hot plate type of kiln

4 Firing with a butane torch

6 Kiln furniture. Nichrome wire mesh supports, asbestos glove and firing fork

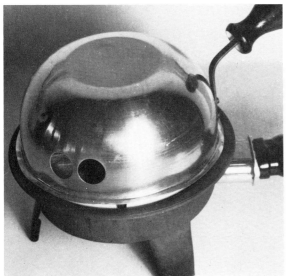

These quantities are for schools; for the home craftsman quarter amounts of these would be sufficient.

These give a variety of colours that may be used for both steel and copper, bearing in mind that if steel is used it must first be given a ground coat of an opaque colour.

In addition to the enamels the following equipment is required:

KILN, enamelling furnace or braising torch.

NICHROME WIRE MESH, firing supports and asbestos slab.

ENAMELLING TONGS, fork or spatula.

METAL SHEARS, pliers, metal punch, steel ruler and graver.

ADHESIVE, ie, gum tragacanth or Polycell.

FINE WIRE (STEEL) WOOL, pumice powder or kitchen cleanser such as *Vim* or *Ajax*, metal file and glass-paper.

FINE SIEVE (nylon tea strainer), brushes or spray for applying gum.

COPPER OR STEEL SHAPES, clean paper, jewellery findings and solder.

VARNISH for protective coating for copper.

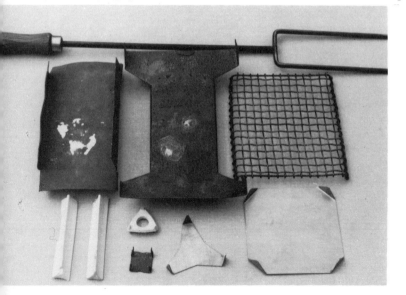

7 Firing equipment. Stilts

Preparation of enamels

Enamel is basically glass fused with various compounds. Lead and borax are used to vary the degree of softness or hardness of the enamel. Alkalis are used to give brilliance and elasticity and metal oxides determine the colour of the enamels.

All enamels have varying degrees of hardness or softness, ie, whether they will fuse at a high or low temperature. Usually transparents are harder than opaques and are less likely to burn out in repeated firings. Red and orange are soft and tend to burn out and so, if used, should be applied last.

Colour tests should always be made before using enamels for the first time. For the opaque enamels tests may be made on pieces of 1 mm (20 g) sheet steel or copper, cut to about 25×35 mm ($1 \times 1\frac{1}{2}$ in.). The steel is then cleaned from grease with acetone or methylated spirits and is then thoroughly cleaned with fine wire wool, water and pumice powder. Dry and then coat thinly with gum tragacanth solution and sift evenly with enamel powder. The enamel powder must completely cover the metal otherwise parts of the steel will show through the enamel as black marks after firing. The layer of enamel should be as thin as possible whilst at the same time giving complete coverage. Dry the piece thoroughly by placing it on top of the furnace. It is important that the glue dries thoroughly as otherwise it would bubble when placed in the furnace, so making the enamel fly off. Place in the furnace when it is bright red, ie, approximately $850°C$ ($1550°F$). If preferred the test piece may be fired by the torch.

At first the enamel will turn black, then become red hot and finally will turn glassy. The enamel has then fused and should be withdrawn from the furnace and put aside to cool slowly in a place free from draughts. Colours change a great deal on cooling and reds and oranges take longest to return to their original colour, changing slowly from black to bright red or orange.

8 Electric kiln, WB2 type. Courtesy of W. G. Ball Ltd

Colour tests for transparent enamels can only be made on copper or tombac (gilding metal). Silver and gold are obviously too expensive for normal school use or for the average home craftsman and so will not be included here. The copper must be cleaned very thoroughly, and there are various ways of doing this. The copper can be immersed in a solution of nitric acid (3 parts water to 1 part nitric acid). *Remember to add the acid to the water.* Instead of nitric acid a 20 per cent solution of sodium bisulphate is equally effective and less drastic on the copper. (See page 39.) However the safest method for school use is to clean the copper with fine steel wool, water and pumice powder or a commercial brand of household cleaner, eg, *Vim, Ajax,* etc. Test to see that the metal is free from grease by holding it under running water. If clean, the water should cling to the whole surface of the metal. If it runs off in globules the metal is still greasy and should be recleaned. Do not touch the surface with the fingers but hold the piece by the edges. Apply gum solution with a soft brush and sift over with transparent enamel. Dry and fire at approximately 820°C (1500°F) until the enamel becomes glasslike in appearance. Flux is often applied first before applying transparents as this gives a clear base.

Transparent enamel can deteriorate if kept for any length of time and for really satisfactory results should be washed before use. Half fill a jar with enamel and fill to the top with water. Pour away the milky liquid which will appear. This process should be repeated several times until the water remains clear. Deposit the remaining enamel paste on to clean paper and dry on top of the furnace or stove. Put back into clean jars in a completely dry state. Powdered enamel is best kept in screw-cap jars so that it is as air-tight as possible. If desired the wet paste may be retained for *cloisonné* or wet application.

If lump enamel is required to be ground, both opaque and transparent enamels should be ground in a mortar with a pestle. The lumps should first be broken into small pieces and covered with water. The pestle should press against the mortar with a rocking action and as the milky residue builds up, pour off and add fresh water. Continue this process until the water is clear and the enamel reduced to the required consistency.

A word of warning. Powdered enamel is equivalent to powdered glass. When sifting, especially with large areas (panels, etc), it is advisable to wear a mask and if children are working with enamels they should be advised against putting their fingers in their mouths.

Preparation of adhesives

Gum tragacanth may be obtained from a chemist in a concentrated solution. This should be thinned to the required consistency (when felt between two fingers it should be slightly tacky) by gradually adding distilled water. It is relatively simple to make the solution if one obtains the gum in powdered form from an enamel supplier. Dissolve half an ounce of gum in a little wood alcohol or methylated spirits. Add gradually one quart of distilled water. A few drops of formaldehyde can be added to the solution to prevent it fermenting.

Polycell wallpaper paste is a good alternative. This can be obtained in small packets, and for a thick solution a teaspoonful of powder should be sprinkled into about a $\frac{1}{2}$ litre ($\frac{3}{4}$ pint) of water. Stir thoroughly and allow to stand for about ten minutes. If too thick, the solution may be thinned with water.

Alternatively glycerine, thinned with a few drops of methylated spirits makes an excellent adhesive.

Pendants and name tags

Pendants or name tags can be made from sheet steel. Use 1 mm (20 gauge) and cut with a metal guillotine. It is easiest to cut rectangular or square shapes and the sizes could be 30 × 40 mm ($1\frac{1}{4} \times 1\frac{3}{4}$ in.) for the pendant and 15 × 60 mm ($\frac{1}{2} \times 2\frac{1}{2}$ in.) for the name tag. All corners and edges must be carefully filed to avoid any jagged or sharp edges. They should then be finished with emery cloth.

A hole needs to be pierced for the jump ring and this should be measured half way along the top edge. Indent the surface with a punch and then drill the hole using a 3 mm ($\frac{1}{8}$ in.) drill. File away any rough edges and clean off any grease with acetone or methylated spirits. Clean thoroughly using fine steel wool, water and pumice powder. Test under running water to see that the steel is free from grease and then dry on top of the furnace.

9 Copper blanks

10 Marking copper prior to cutting

11 Punching hole with a screw punch

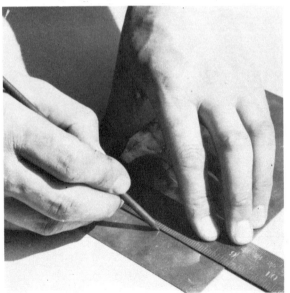

Copper and glass panel. Group project by 12-year-old boys

12 Filing the edges
13 Cleaning with steel wool
14 Painting with gum tragacanth
15 Sifting with enamel

When cool, coat with gum tragacanth and apply this with a camel-hair brush. Sift over very evenly the first coat of opaque enamel. The sifter should be held approximately 300 mm (12 in.) above the article and shaken gently. Lift the piece and wipe the edges free from enamel with the side of the finger. Blow off any loose enamel and clear the hole with a steel pointer. Place the piece on a *Nichrome* wire mesh trivet and, with the aid of firing tongs or spatula, place in the furnace (820°C, 1550°F).

When the enamel turns glassy remove from the furnace. If on removal the surface appears bumpy, like the peel of an orange, it is slightly underfired. Put back into the furnace for a few seconds until the surface becomes smooth and glasslike. Remove and leave to cool.

Take a few small pieces of opaque enamel lumps, coat the pendant with gum and put the pieces on top. When the gum has dried, refire the pendant until the enamel pieces or

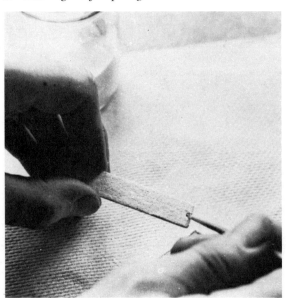

16 Cleaning the jump ring hole

17 Firing finished pendant

'jewels' have melted into flat pools, then remove from the furnace.

Much of the success of a simple pendant using this technique depends on the colours chosen. For instance if the ground colour was black, white or bright-coloured opaque pieces should be used. As the back of this piece has not been counter-enamelled, the loose firescale* on the back should be removed with emery paper or a wire brush. The back may then be sprayed or painted with silver, gold or copper paint. A jump ring is fixed through the hole and leather thonging or a chain attached.

For the name tag, the steel is prepared in the same way and the first coat of enamel applied and fired. When cool, the piece should again be coated with gum and a contrasting opaque colour sifted on top, eg, ground coat black, second

*Firescale is a black oxide which forms on copper when heated.

18 Pendant with lump enamel

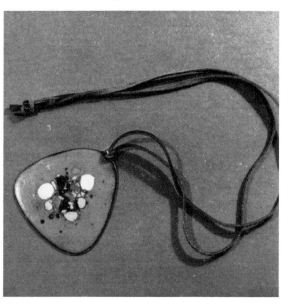

19 Various sgraffito pendants

coat bright red. Wipe the enamel from the edges and clear the hole. With a sharp-pointed instrument scratch through the top coat the letters that are required for the name. This takes a little practice and it is a good idea to write the name on a piece of paper the same size as the metal tag. The letters can then be spaced out evenly. If a very fine pointer is used at first, eg, a needle, the letters may be scratched through afterwards with a heavier instrument. Blow away any surplus enamel and fire. This process of scratching through a top layer to expose the under surface is called *sgraffito*.

20 Pendant with fired mosaic pieces

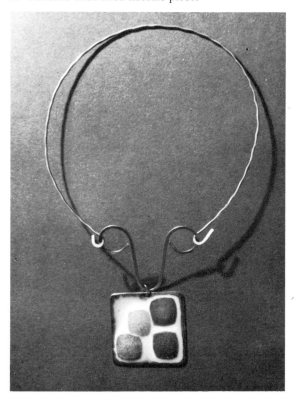

21 Champlevé ring, French

Masking solution
Copper protector

In order to prevent firescale building up on the back of a copper piece which has not been counter enamelled, eg, jewellery, dishes, etc, a protective coating may be applied. If difficult to obtain from a supplier it is possible to make up a satisfactory solution with simple ingredients. You will need some clay slip (use ball clay which is more plastic) and mix with water until it is of the consistency of thick cream. Add to this a few drops of *Teepol* (U.K.), *Duponal* or *Synthrapol* (U.S.A.) (a concentrated industrial detergent) and a few drops of methylated spirits. Shake these well together and keep in a tightly screw capped jar. The mixture should be painted evenly over the back of the article and left to dry on the top of the furnace. The front surface of the copper may then be coated with gum and enamel and fired in the usual way. On removal from the kiln, the protective coating will fall away from the metal, on cooling, leaving the copper perfectly clean.

For repeated firings the back of the article must be re-coated with this solution each time before refiring.

Swirling

Attractive pendants can easily be achieved by swirling different coloured enamels when still in a molten state.

A swirling stick can be made from a mild steel rod, 3 mm ($\frac{1}{8}$ in.) diameter. Cut a piece approximately 450 mm (18 in.) in length and bend 10 mm ($\frac{1}{2}$ in.) from the end at a right angle. File or hammer the end into a point. Fix the other end into a wooden handle.

The pendant must be given a base coat of enamel and then fired. When cool, coat with gum and place a few pieces of contrasting coloured enamels on top. When the pieces have melted into flat pools, open the furnace and place the pointed end of the swirling stick on the surface of the enamel. Rotate the end gently with a swirling motion so that the enamels are stirred together in a molten state. Withdraw the rod and leave the pendant for a few seconds longer in the

24 Swirling
23 Second firing with lump enamels

22 Firing the ground coat

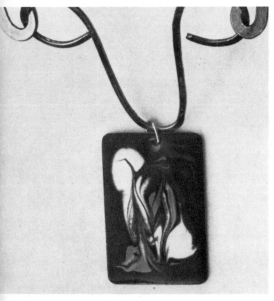

furnace. Withdraw and allow to cool. If a hot-plate type of kiln is being used the cover should not be removed but the stick should be placed through one of the holes in the lid for the swirling. This is because the heat loss is too great to melt the pieces if the cover is removed.

The swirling stick should not be pressed too heavily into the enamel as otherwise it will expose the metal surface. Much of the success of swirling depends on the choice of colours used. If too many different coloured pieces of enamel are mixed together in one piece the result will look vulgar. It is best to try subtle colour combinations together of not more than two or three colours.

25 Finished pendant
26 Showing how to swirl with a hot-plate type of kiln
27 Swirling with lid removed, to show technique

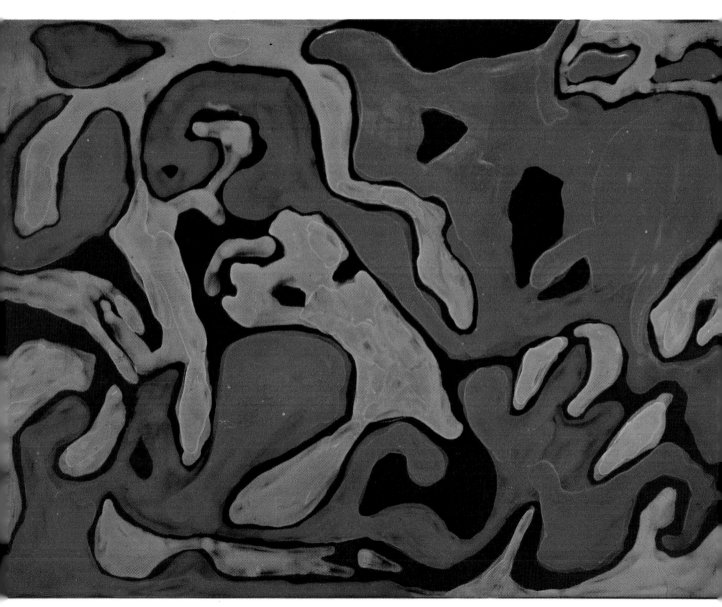

Steel panel. Slush (painted) enamel, by the author

Leaf stencil pendant

Using a steel ruler and graver mark off the length of the pendant required on 1 mm (20 g) copper sheet. Cut the copper with a guillotine, or metal shears. File the edges and corners carefully and finish with fine-grade glass-paper. Make a hole for the jump ring by using a 2 mm ($\frac{1}{16}$ in.) drill or metal punch. File away any rough edges, then clean the metal with fine wire wool (grade 00) and pumice powder. Test under running water to ensure that the metal is free from grease. Coat with gum tragacanth and dust on the first coat of enamel, eg, transparent flux. Pour back the residue of the enamel into the jar. Dry on top of the furnace until the gum has completely dried. Fire in the furnace at approx. 820°C (1500°F) until the enamel melts and appears glasslike. When cool, re-apply the gum and place the stencil (ie, a leaf) on the top surface of the pendant. Re-dust with coloured transparent enamel. Using tweezers, carefully remove the stencil (leaf) and add small pieces of transparent lumps. Fire until the enamel pieces melt into flat pools.

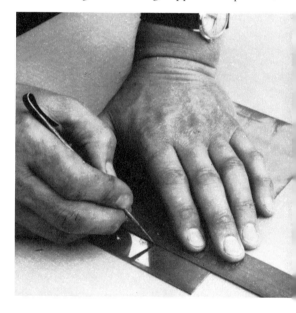

28 Measuring and marking copper for a pendant

29 Punching a hole

30 Fixing leaf stencil with gum on to enamelled pendant

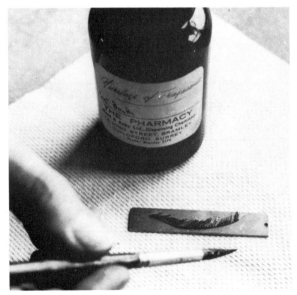

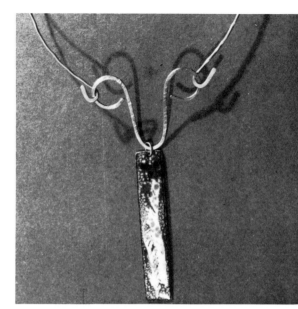

33 Finished pendant

31 Removing leaf after dusting with enamel

32 Placing lump enamel prior to firing

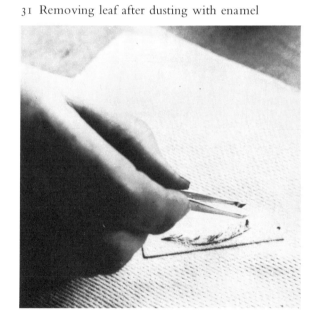

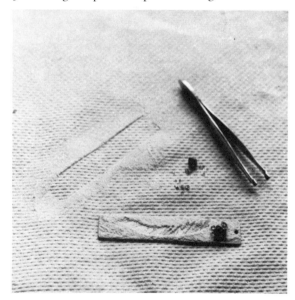

Counter enamelling

This means applying a coat of enamel to the back of an article and is necessary if *cloisonné* or lump enamel is to be used as the form of decoration. This is because stresses, set up during the cooling of the metal, would cause the lumps or *cloisons* (unless previously braised to the metal) to crack away from the piece.

If copper is used the counter enamel, or underside, can be applied first. In this case the top surface would have some firescale fixed to it on removal from the kiln. This must be removed and the top surface thoroughly cleaned before the first application of enamel. Alternatively, if the dusting or sifting technique is used, the enamel may be applied to both surfaces and fired simultaneously. In this case special supports must be made to support the piece at the edges only. Personally I find it easier to use a slush (liquid) enamel for the counter

34 Applying liquid counter enamel
35 Coating the front with copper protector

36 Copper protector after firing

36 (*a*) Pendant with lump enamel

enamel and to use the dusting technique for the top surface. The liquid enamel may be applied with a soft brush to small areas but should be sprayed evenly on to larger surfaces. The article must then be dried thoroughly before firing.

When counter enamelling takes place the firing supports are most important. They must be made especially for the article so that it is supported at the edges only or have very fine steel stilts so that the minimum surface or enamel is in contact. After firing and cooling the stilt can be knocked away from the piece. The counter enamel must not be applied too thickly as otherwise the stilt will become too firmly embedded in the enamel. If preferred, a small mask may be placed where the stilt would come, so leaving the metal bare. The mask must be removed before firing and the metal patch cleaned afterwards. Masking is only practical when certain articles are counter enamelled, eg, dishes or brooches where the pin would be soldered on to the masked area.

Cloisonné

Cut and clean the copper piece. Pierce a hole for the jump ring if the finished article is to be a pendant. Bend fine copper or silver wire to fit the shape of the pendant and flatten the wire with a mallet or hammer. Apply the counter enamel to the back of the piece and a coat of masking solution to the front surface. After firing clean the top surface of the pendant. Apply gum to the top surface and sift with the first coat of enamel, eg, transparent flux. Carefully place the bent wire or wires on top of the powdered enamel. Dry thoroughly on top of the furnace, then fire. After firing, clean the copper wire with a carborundum stick and water, using a rotary action. Add gum tragacanth with an eye dropper to finely ground enamel. Mix it to a smooth paste and apply it between the wires by using a spatula and pointer or other suitable tool. Allow to dry thoroughly before refiring. After firing, again clean the copper wire with the carborundum stick. If necessary refill the areas with enamel paste and refire. Instead of the wire a paper clip may be used as an experiment.

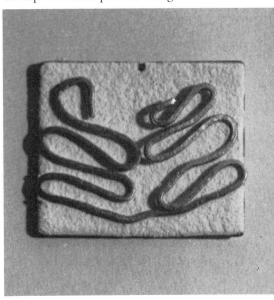

37 *Cloisonné*. Shaped copper wire placed on top of enamel prior to firing

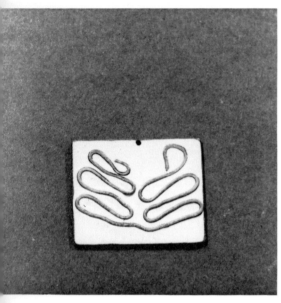

38 After the first firing
39 Stoning with carborundum stone

40 Dropping gum on to powdered enamel

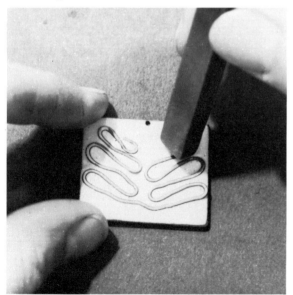

41 Mixing enamel into a paste with gum

42 Filling the *cloisons* with paste enamel

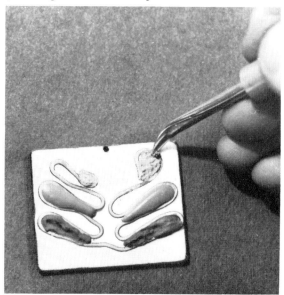

A *cloisonné* pendant using silver *cloisons*

Design the shape of a bird on paper and cut it out to form a template. Mark round the template on the copper sheet and cut carefully with shears. File any rough edges and clean thoroughly. Apply the counter enamel to the back and cover the front surface with a protective coating (masking solution). With the top surface face down on the *Nichrome* wire mesh put into the furnace. After removal, the protective

43 Bird pendant. Copper blank

44 Applying the counter enamel

coating will fall off as the metal cools. Re-clean the front surface and dry, then coat with gum and sift over with transparent enamel or flux. Position the cloisons, made from silver wire, where required. Place the pendant on a metal stilt on to the trivet and insert into the furnace and fire until the enamel turns glassy. Remove and allow to cool. Mix transparent or opaque enamels to a paste and fill the rings, or *cloisons*, with the colours. Dry thoroughly and re-fire.

If necessary the silver *cloisons* may be cleaned by brushing with a glass fibre brush or silver polish.

45 Painting the front with copper protector

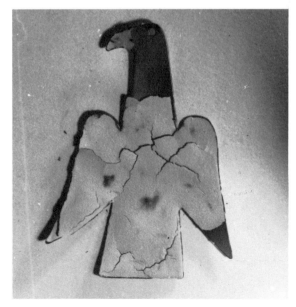

46 Copper protector after the first firing

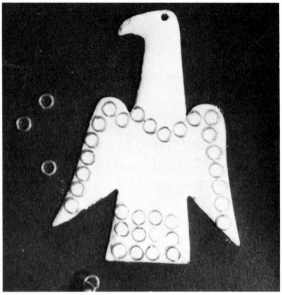

47 Placing silver *cloisons* on to powdered base coat
48 After first firing
49 Filling the *cloisons* with paste enamel
50 Finished pendant

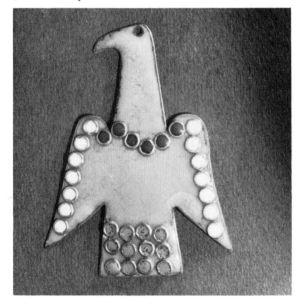

Enamelled copper and silver brooch

Silver wire may be used to enhance enamelled pendants, brooches, etc. A short length of 1·62 mm silver wire may still be obtained with little expense. The wire should be cut into small pieces, 10 to 25 mm ($\frac{1}{2}$ to 1 in.) in length, and flattened with a planishing hammer. The wire may also be cut into 5 mm ($\frac{1}{4}$ in.) pieces and formed into flattened balls by placing on a charcoal block and heating with a torch. Heat until the silver melts and runs into a ball shape. The silver may be cleaned afterwards by immersing in a bright dip (equal parts of nitric and sulphuric acid) or safer to use a proprietory brand of silver cleaner.

The copper sheet should then be cut to the required shape. If it is to be made into a brooch a paper mask should be cut the same size as the back of the brooch mount. Attach the mask to the back of the brooch and paint over the entire surface with liquid enamel. Allow to dry, then carefully remove the paper mask. Turn over on to the front surface and paint with gum tragacanth. Dust over with transparent enamel or flux, then put the piece on to a stilt and fire. Re-coat with transparent enamel and position the silver pieces on the powdered enamel surface. Again put the brooch on a steel stilt and fire. After removal from the furnace the silver may be cleaned with silver polish or silver dip. Clean the exposed area of copper on the back of the brooch and soft solder the brooch pin in place.

51 Melting silver wire on a charcoal block

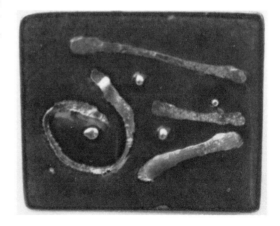

52 Enamelled brooch decorated with silver wire

Enamel and copper wire brooch

Copper wire may be cut and bent to form a linear structure on the surface of a brooch or pendant. Design and cut a shape from sheet copper 1 mm (20 g). Snip 1 mm (20 g) wire to required lengths and bend to fit the shape of the piece. File the edges and clean the copper then paint with gum and sift over with powdered enamel. With the aid of tweezers carefully put the pieces of wire onto the copper shape. Insert into the furnace and fire. Clean any firescale from the back and copper wires by immersing in the dilute acid solution. When clean, solder the brooch pin onto the back.

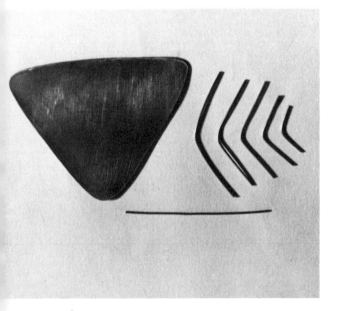

53 Copper blank with copper wires

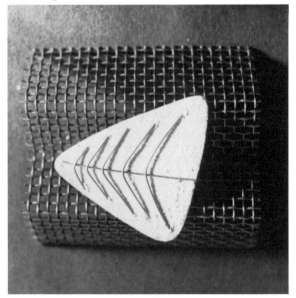

54 Wires placed on to powdered enamel base ready for firing

55 The finished brooch

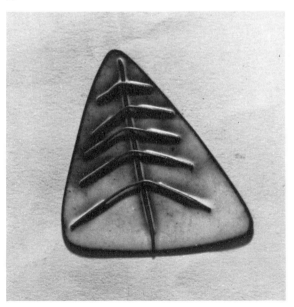

56 Pendant decorated with wire and crystal pearls

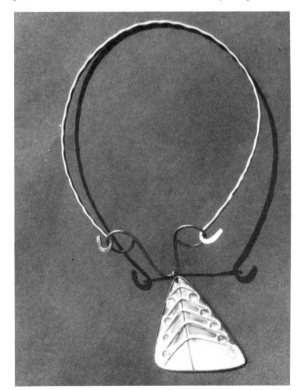

Pliqué à jour pendant and earrings

Cut the shapes for the pendant and earrings by using a paper template. File and clean the edges of the copper and pierce holes for the jump ring and earring findings. Using various size drills (not larger than 5 mm, $\frac{3}{16}$ in. in diameter), pierce holes right through the metal. (I used two drills of 3 and 2 mm, $\frac{1}{8}$, $\frac{1}{16}$ in, diameter.) Coat the copper with the masking solution. This prevents the firescale building up on the copper surface. Put the pieces on to a sheet of mica. Meanwhile mix finely ground transparent enamels to a paste with gum tragacanth. Insert this enamel paste into the holes until the area is level with the copper. Leave on top of the furnace to dry out thoroughly. Put the mica with the pieces on it on to a trivet and fire. After removal from the furnace the firescale, if any, must be cleaned from the copper. Re-coat with the protective solution and replace on to the mica sheet. The enamel will have sunk below the level of the copper and will require refilling with enamel paste. Again dry thoroughly and re-fire. Continue in this manner until the areas are filled level with the copper surface. The copper may then be given a final clean and polish. When held to the light the holes, filled with transparent enamel, will glow like small stained-glass windows.

It is possible to obtain copper strips with small circles and squares already cut out of the metal. These strips could then be cut to the required size to form jewellery pieces or mobiles.

57 *Pliqué à jour* earrings and pendant

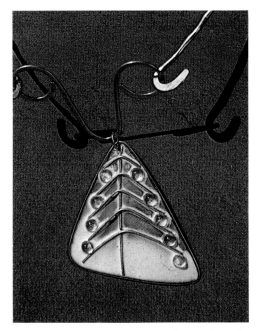

Pendant, decorated with wire and crystal pearls, by the author

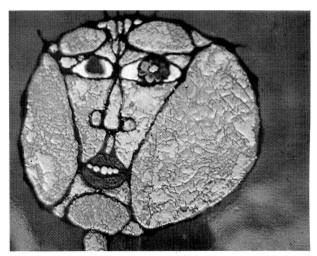

'Head' Copper panel by Gerda Flockinger

'*Frost*' completed panel

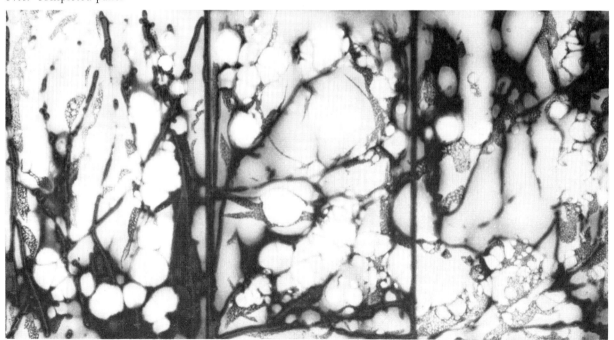

Metal foils

Silver or gold foil (leaf) is obtainable from suppliers (see page 87). Foils are usually sold in sheets 100 × 100 mm (4 × 4 in.), separated by sheets of thin paper. When cutting foil the sheets of paper should be retained so that no grease is deposited on the surface of the foil.

There are various ways of using foil to decorate or enhance enamel and they may be used under transparents, so giving an appearance of silver or gold instead of copper. A ground coat must always be first applied to the copper, usually of transparent flux, but for some effects, transparent, opaque or opalescent enamels may be used. Before the foil is applied to the base coat it must first be pricked over with a steel needle to enable any gases to escape during the firing. A dressmaker's steel spiked wheel is excellent for the purpose. Always retain the foil between its protective paper and press

58 Silver foil, dressmaker's tracing wheel and scissors

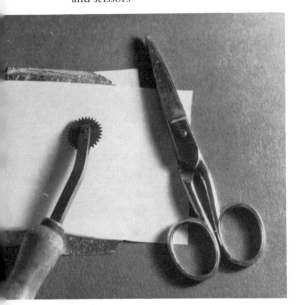

59 Punching shapes in silver foil

the wheel across it several times. The protective paper should then be lifted off and the foil, with a gum-moistened brush, placed on the gummed surface of the enamel. Allow the gum to dry out completely before firing. The piece should then be inserted into the furnace and removed after a few moments before complete fusion has taken place. The foil must be burnished with the end of a knife or spatula and then re-inserted in the kiln until fusion is complete.

When cool the piece may be decorated in several ways:

(a) Wet application with transparents and opaques.

(b) Sifted over with transparent enamel. Before firing a sgraffito (to scratch through the top layer to expose the underneath surface) design can be made to mark a silver line showing through the enamel.

(c) Small pieces of transparent lump enamel may be placed on the surface so that the foil will show through from underneath.

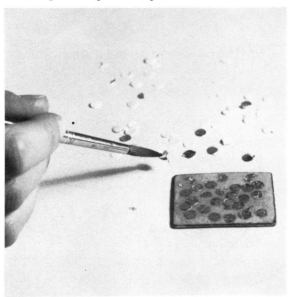

60 Fixing foil shapes on to pre-enamelled base

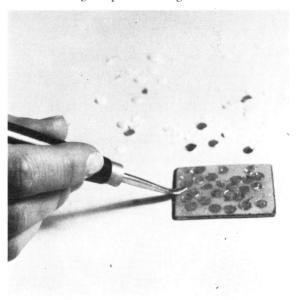

61 Burnishing foil prior to firing

(d) A design can be built up using stencils, transparents and foil cut to various shapes and sizes. With several thin layers of transparents and foils this can achieve an almost three-dimensional effect.

(e) The foil can be cut into small squares or strips, punched into confetti-like pieces and used as decoration underneath transparent enamels or transparent lumps or glass pieces.

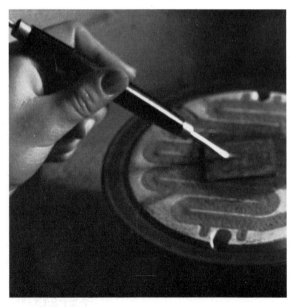

62 Burnishing foil during firing

63 Finished brooch showing silver foil shapes

Wet application of enamel

As an alternative to sifting the enamel (as a dry powder) on to the metal, the enamel can be applied in a liquid state. This enables several colours to be applied at the same time and fired together. A ground coat of transparent flux should be sifted on to the cleaned copper and fired. The colours, transparents, opaques or both, should be finely ground with distilled water. A thin glue solution may be added to form a liquid paste. The colours should be applied with a spatula or very fine brush and placed alongside each other to form the design. The piece should then be placed on top of the furnace to dry off any moisture. When the enamel is completely dry it should be fired in the usual way.

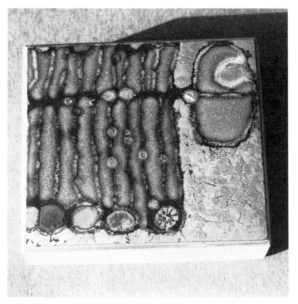

64 Wet application of enamel with gold lustre. Small panel by Gerda Flockinger

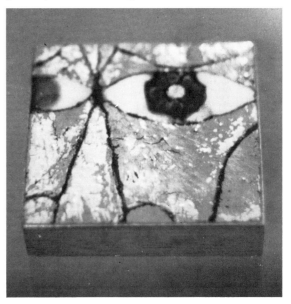

65 Wet application of enamel with silver foil. Small panel by Gerda Flockinger

Lustres

Lustres may be applied for the final firing if desired. They are always applied last as lustres fire at a lower temperature than enamels. Paint the lustre on the article with a soft camel-hair brush, or, for fine detail, use a pen. Allow to dry as thoroughly as possible. As metallic lustres are suspended in oils, this oil must be burnt away with the door of the furnace open.

Insert the piece, when dry, in the kiln at a lower temperature (750°C, 1350°F). Keep the door of the kiln open to enable the fumes from the oil to escape. When the oil has burnt away, close the door and observe continuously through the spy hole. The pyrometer (a temperature gauge attached to kilns) should read 750°C and the lustre will appear shining and metallic. Remove the piece from the furnace.

If lustres are overfired they will burn away. If this occurs the lustre must be re-applied and the same procedure repeated.

Making a copper dish

Design the shape of the dish on paper and cut out to form the template. Using a steel graver mark round the template on the copper sheet, then cut round with metal shears. Shape the bowl by using a leather mallet and hollowed wooden block. File the edges and finish with glass-paper. Clean the copper surface thoroughly with fine steel wool, water and pumice powder. Spray or paint with gum tragacanth and sift on the first coat of enamel or flux. Dry thoroughly and fire. Enamel colours very easily pick up tiny particles of firescale. These will fire on to the enamel as black particles, so it is very important to keep all enamels scrupulously clean.

Once a single colour has been achieved on a dish many different techniques may be tried.

(a) Sgraffito. This is drawing or scratching through a layer of powdered enamel to the already fired base coat.

66 Marking round the template

67 Cutting out with metal shears

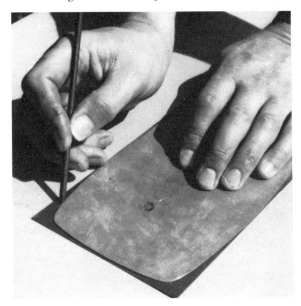
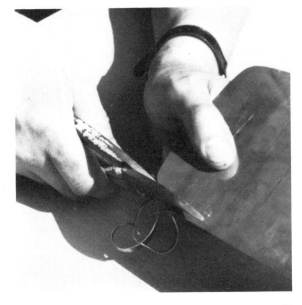

68 Beating dish from a wooden mould

69 Filing edges of the dish

70 Cleaning copper with steel wool

71 Base coat

72 Firing dish

(b) Liquid enamel or gum may be applied freely with a brush to form a design on a once fired base coat, and pieces of enamel can be added to make pools of colour when melted.

(c) Mosaic. Small pieces of lump enamel may be placed next to each other to form a mosaic on top of a pre-fired base coat and re-fired.

(d) Stencils. Various stencils can be cut from paper, or made from leaves, dried grasses or flowers, and enamel sifted over and the stencil removed before firing. Stencil designs can be built up in layers by using transparent enamels, each colour having a separate firing.

(e) Nets. Various meshes, string, threads or nets may be used in the place of stencils.

(f) Wet application of enamel may be applied with a spatula or brush to form a design of several colours placed next to each other and fired together in a single firing.

(g) Silver and gold foils may be used under transparent enamel to make free-form or hard-edge designs.

(h) Lustres may be added (sparingly) to accent certain areas of the design or drawn with a pen on a pre-enamelled base.

(i) Transparent enamel sifted on to a pre-fired opaque (soft) enamel and fired at a high temperature will produce a mottled effect due to the soft opaque fusing first and so breaking through the surface of the transparent enamel.

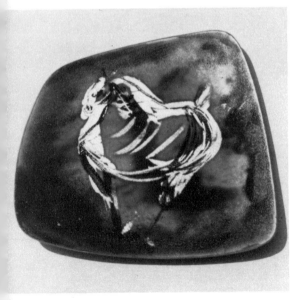

73 Free form design

74 Dish showing sgraffito design. Courtesy of COID
75 Dish showing sgraffito design

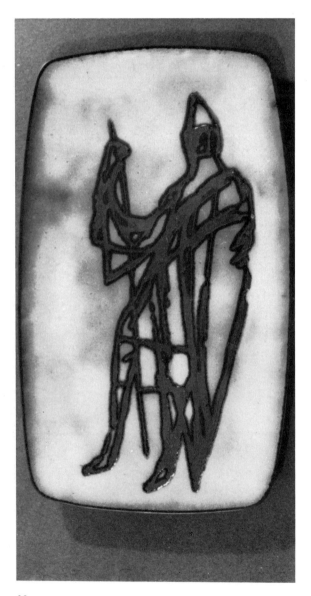
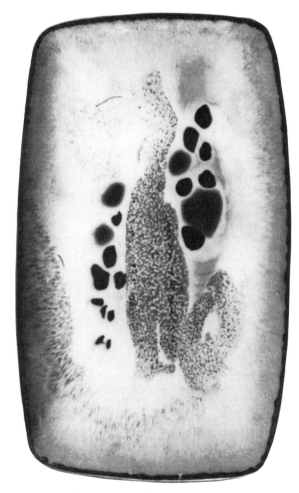

76 Dish showing sgraffito design
77 Free form design with lump enamel

78 Free form design
79 Free form design with lump enamel

80 Opaque enamel painted freely over transparent with one transparent piece. Courtesy of COID

81 Dish from Limoges

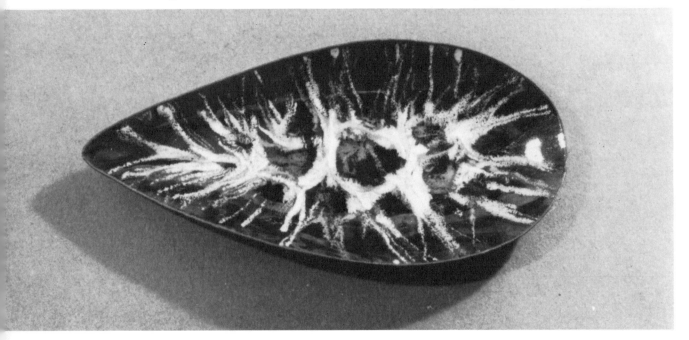

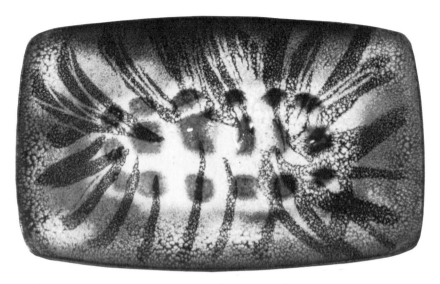

82 Free form design with transparent lump enamel
83 Long dish. Free form design with pieces

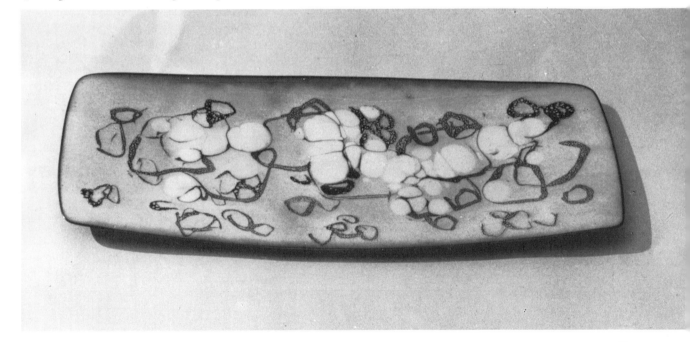

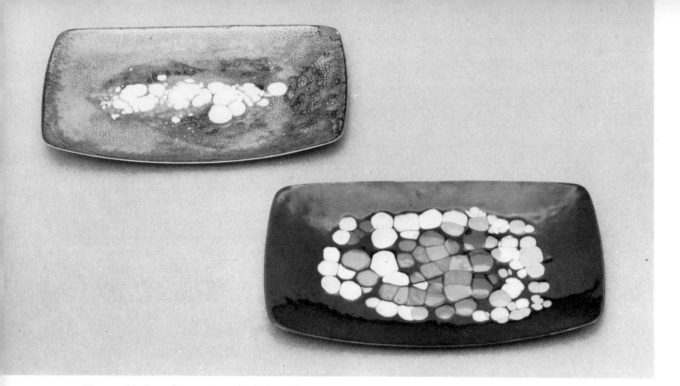

84 Pieces added to form a mosaic design. Courtesy of COID

85 Mosaic designs with lump enamel

86 Cigarette box. Mosaic design with lump enamels

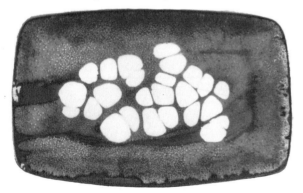

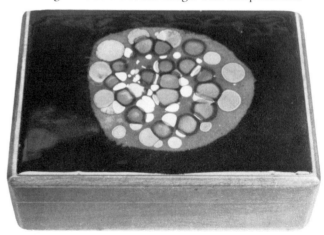

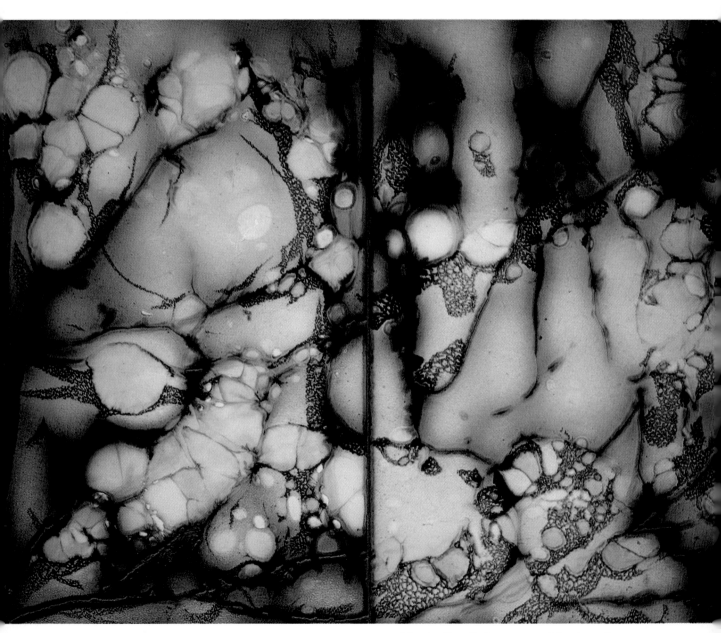

'Frost'   Detail of copper panel by the author

87 Cutting newspaper for stencil strips

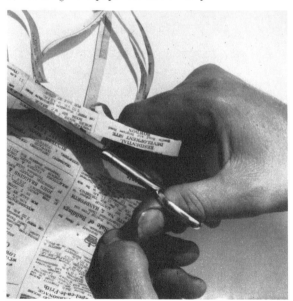

88 Gluing strips on to pre-enamelled base

89 Sifting over a layer of powdered enamel

90 Removing the stencil

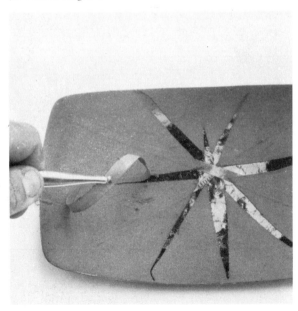

91 Prior to second firing

92 Finished dish

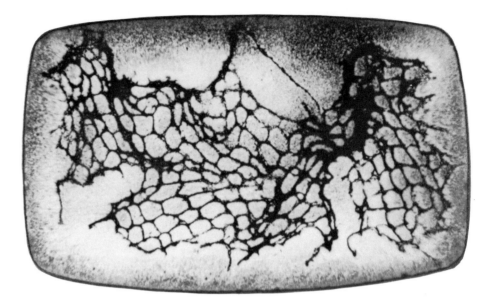

93 Dish showing net stencil. Courtesy of COID

94 Transparent enamel over opaque with lump enamel fired at a high temperature

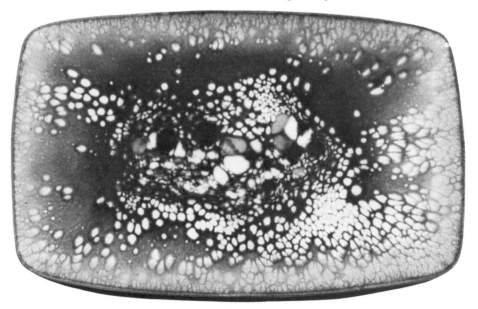

95 Cigarette boxes. Sgraffito and net stencil designs

The use of firescale for decoration

Firescale is a black oxide which forms on copper when it is subjected to the heat of the furnace or braising torch. It may be used to form the basis of a design.

The copper is inserted into the furnace or heated with a propane torch until the oxide forms on the surface of the metal. When cool, brush off any loose firescale with a wire brush. The surface is then coated with gum and sifted over with transparent flux or coloured transparent enamel.

96 Ashtray showing firescale decoration

If using transparent flux (soft fusing) the copper oxide can be made a rich orange colour by altering the firing temperature. The piece should first be inserted at a high temperature (approximately 900°C, 1650°F). After withdrawal and cooling it is then re-inserted at a lower temperature (820°C, 1500°F) and withdrawn after about one minute and prior to complete fusion taking place.

If preferred the design can be controlled by first applying enamel (transparent or opaque) to certain areas of the copper surface. After firing the firescale will have formed on the exposed parts of the copper not covered with enamel. The whole surface should then be covered with transparent flux or transparent enamel.

97 Free form design over firescale with lump enamel

Transfer designs *Printing*

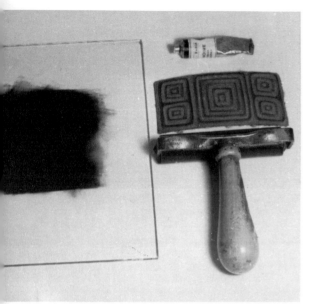

Any raised or engraved surface may be transferred to a pre-enamelled surface, eg, linocuts, etching plates, wood grain surfaces, carved plaster slabs, fired clay tiles, polystyrene, etc.

The enamelled surface should be cleaned with detergent to remove any grease. Squeeze some lino ink or other oil-bound printing ink on to a slab of glass and roll out with a rubber roller; then ink the printing block with the roller. Take a piece of waxed or tracing paper and place over the block. Rub the surface firmly so that the ink is evenly transferred to the surface, then press this print on to the enamelled plate. Remove the paper, dust over with finely ground enamel and fire in the usual way. The same design may be repeated many times by using this method.

98 Materials required for lino printing
99 Copper blank ready for sifting with enamel
100 Inked copper blank ready for sifting with enamel

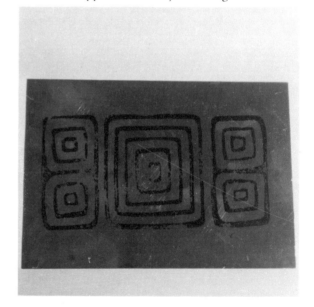

If a linocut is used, the block is inked and may then be pressed directly on to the pre-enamelled surface. After removal of the block, enamel powder is sifted over, any surplus enamel blown away and the piece then fired in the usual way.

Similarly leaves, grasses, etc, can be inked over with the roller then pressed on to a pre-enamelled surface. When the leaf is carefully removed powdered enamel is sifted over the inked imprint and then fired in the usual way.

Direct printing can be achieved in the same way by using sponges (both synthetic and natural) and even potato cuts. Press the sponge or potato on to the inked slab and then press on to the pre-enamelled base. Sift over very finely ground enamel (100 to 150 mesh).

101 Finished cigarette box

Silk screen printing

If several articles are required using the same design, silk screen printing can provide the answer. Opaque enamel should be used on a pre-enamelled ground.

It is possible to purchase silk screens in a wide range of sizes, but it is comparatively easy to make a small screen oneself. The materials required are as follows: wood strip (38 × 15 mm, $1\frac{1}{2} \times \frac{5}{8}$ in.), organdie, silk, *Terylene (Dacron)* or nylon, drawing pins, masking tape and a length of rubber strip for the squeegee.

Make a frame to the required size with the wood strip. The size of the interior of the one illustrated is 150 × 200 mm (6 × $7\frac{3}{4}$ in.). The corners of the frame should be mitred. Take a piece of organdie 25 to 50 mm (1 to 2 in.) larger all round than the outside measurement of the frame. Stretch this over the frame and keep in place with the drawing pins. Stick

102 Silk screen equipment

103 Pushing slush enamel through silk screen

masking tape all round the inner edge of the frame, so that it masks off approximately a 25 mm (1 in.) border of the organdie. As the material is transparent it may be placed over the design and traced with a pencil or ballpoint pen. All areas not required to print must then be stopped out. This can be done by using a stopping-out varnish, a polymer medium such as *Marvin Medium* (U.K.), *Sobo* or *Elmers glue* (U.S.A.) or *Profilm* (U.K.) or *Nu-film* (U.S.A.). Profilm is a transparent film which is easily cut with a stencil knife and then attached to the surface of the screen by ironing.

When dry, the frame is ready for printing. The article (this should be flat) must be pre-enamelled and placed in position under the screen. Special enamel for silk screen printing may be obtained from enamel manufacturers or it can be applied in the following manner. Gum tragacanth is made up into a thick jelly and pushed through the screen on to the enamelled base. The screen is then lifted carefully and finely powdered enamel (150 mesh) sifted over the gum. Shake away any surplus enamel powder and dry thoroughly. Fire in the usual way.

The squeegee may be made from a piece of rubber strip—draught excluder, windshield wiper, etc—and inserted in a piece of wood the same width as the interior of the frame. The enamel is pressed through the areas of the organdie not stopped out, and so is transferred on to the pre-enamelled surface.

104 Silk screened enamel panel

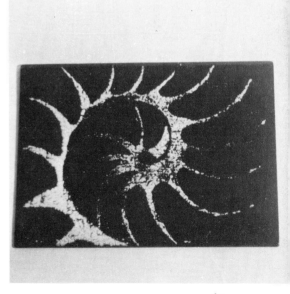

Group projects for schools

1 An enamelled steel panel

A simple project for group work is a panel composed of steel tiles of varying sizes and enamelled in a limited number of colours. It is wise to limit the number of colours to be used as otherwise the panel can end up looking like a patchwork quilt. As steel is the metal to be used the enamel colours must be opaque and not transparent.

The design should be worked out to scale on paper. The steel tiles can then be cut accurately with a metal guillotine. Each child can be responsible for enamelling one tile.

Opaque enamels supplied for copper do not always take successfully on steel, and for a large panel it is advisable to obtain enamels especially prepared for steel. However, many of the opaque enamels for copper do fire successfully on steel and one should first make colour tests to find out which colours fire well.

If firing larger panels, after degreasing, they should be pickled in a 10 per cent solution of hydrochloric acid. After thoroughly rinsing in water, dip them in a bowl or bucket of liquid ground coat enamel and allow to drain. When dry fire on stilts at a temperature of 840°C (1540°F). The panel is then ready for a base coat of liquid white or black enamel. When dry, fire at approximately 820°C (1500°F).

Enamels for sheet steels may be obtained from suppliers (see page 86). They are usually supplied in powder form and should be mixed with water until of the consistency of clay slip. Pass through a fine sieve to get rid of any lumps.

2 Copper and glass panel

Small trays of copper can be made in varying dimensions from 25×40 mm to 50×50 mm ($1 \times 1\frac{1}{2}$ to 2×2 in.), the edges being raised 5 mm ($\frac{1}{4}$ in.). The edges must be filed and the backs of the trays made level. Clean thoroughly with

105 Enamelled steel panel (sgraffito). Boys 13 years. George Abbot Boys' School, Guildford

pumice and wire wool. The backs of the trays must be counter enamelled because lump or glass is to be fused to the top surface. After counter enamelling immerse the trays in a 20 per cent solution of sodium bisulphate (mix one part to five parts of water) in order to remove the firescale from the inside of the trays. When clean (the copper will look pink) remove with brass tongs and rinse thoroughly under running water. Clean the front surface again and coat with gum then sift over with flux or transparent enamels in the required colours. As each piece has been counter enamelled it must be placed on steel stilts in the furnace when firing.

After firing, assemble all the pieces together to see which colours need emphasising. Take pieces of coloured glass, mosaic or lump enamel and place on the trays where required. Fix into place with the gum solution, dry and fire. If necessary, re-clean in the acid solution, otherwise clean the edges of the trays with a carborundum stick. The pieces may then be assembled and mounted on to a wood panel.

Enamelled mobile

Design shapes on paper and cut out to form templates. Place the templates on sheet copper (0·70 mm, 22 g) and mark round with a scriber. Cut out the shapes with metal shears and drill holes for the jump rings or connecting links. File all the edges carefully and finish with fine grade glass paper. Pickle the copper in a 20 per cent solution of sodium bi-sulphate or clean with fine steel wool and pumice powder. All pieces are enamelled with the required colours; simple, bright, opaque colours being perhaps the most suitable. When all the pieces have been enamelled lay them out for assembly. The wire structure must be strong enough to support the suspended enamel pieces. The balance may be adjusted as it is put together.

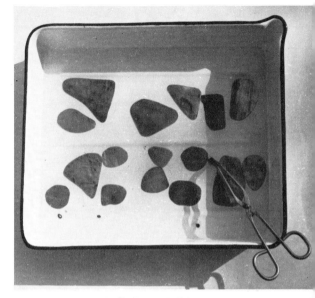

108 Cleaning copper pieces in diluted acid

106 Copper shapes for a mobile

107 Drilling hole for connecting ring

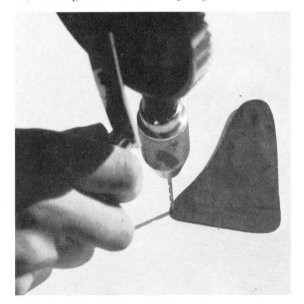

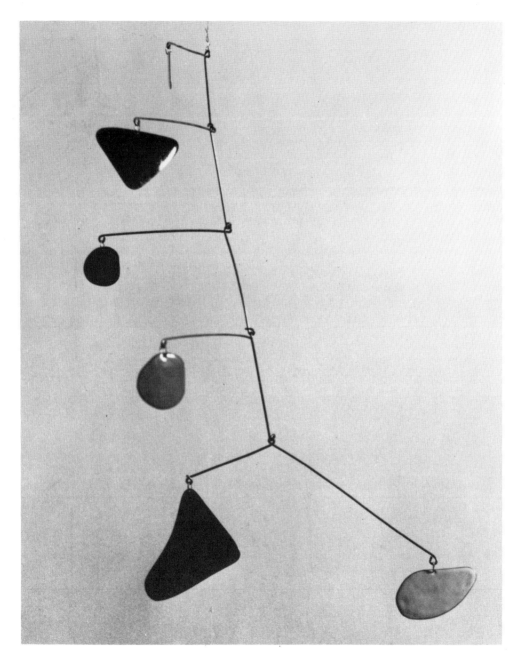

109 The assembled mobile

Sculpture and three-dimensional panels

Some sculptures may be enamelled, eg, steel constructions, copper (silver soldered) and to some extent welded steel. The stilts or supports usually require to be made especially in order to support the piece at the edges. Three-dimensional panels may also be produced, being assembled with screws or nuts and bolts, or by an epoxy resin, impact glue, or mounting at different levels.

Enamelled knight (chess piece)

This was conceived as a flat section, cut out and then bent to form the figure. After drawing the shape cut a template in card or aluminium. Place this on a sheet of 1 mm (20 g) copper and mark round with a scriber. Cut out using metal shears. Bend or hammer carefully over a rounded piece of wood, eg, a broom handle. The copper must be cleaned by

112 Knight after dipping in slush enamel

110 Template of knight prior to forming

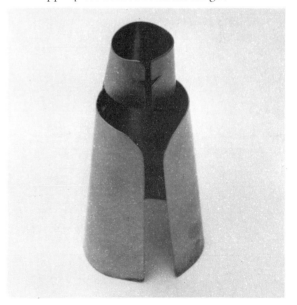

111 Copper piece bent to form the knight

113 Finished piece showing gold lustre after final firing

inserting into a dilute solution of nitric acid or sodium bi-sulphate and rinsed under running water.

It is not practical to enamel sculpture by the dusting process, and to obtain the most satisfactory results the piece should be dipped or sprayed with liquid enamel. The enamel should fill a container deep enough to hold the sculpture suspended on a wire. It is often necessary to dip the sculpture several times in order to build up a sufficient thickness of enamel. After the final application and when completely dry, lay the piece on the *Nichrome* support, taking care that the edges only are touching the support. Fire in the usual way.

A chess set could be produced by using this method of one piece of metal. The pieces are made by making slight alterations in the design and cutting templates for any piece requiring repetition.

114 Detail of panel. Copper shaped with oxyacetylene welding torch and enamelled with transparent enamel. *Courtesy of Whibley Gallery*

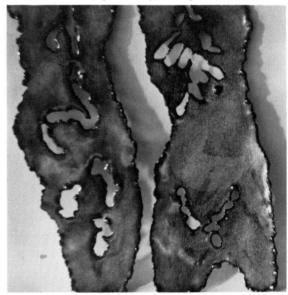

115 Enamelled welded steel sculpture. Paul Barter

Colour and design

I have left this subject until last because it is necessary to understand the craft of enamelling before beginning to design or decorate a piece of work.

Design is a personal matter mainly dictated by individual tastes. Enamelling is an exacting craft: once fired, it is not possible to eradicate or change the colour and design as one can in drawing and painting. To avoid disappointments and a waste of materials it is preferable to make preparatory designs. These may vary from an exact drawing to scale to rough colour notes or quick sketches.

When decorating a dish or other specific piece, one must bear in mind that the design should fit the shape that it is to decorate. The shape of the piece should be drawn several times on paper and the deigns can then be worked quite

116 Copper panel using firescale for the basis of design. *Planets*.

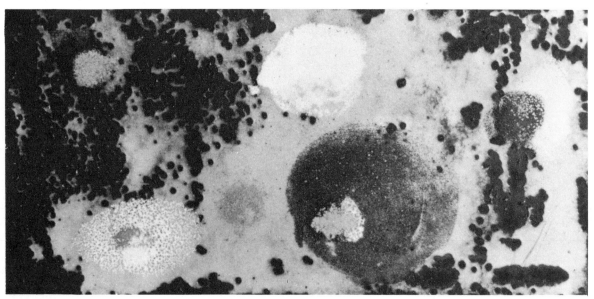

freely on the drawn shape. Decide the method of decoration to be used, eg, sgraffito, foil and transparents, wet application, mosaic, stencils or print transfer. The design must obviously be adapted to the kind of technique to be used and to the shape of the piece to be decorated. It is also a good idea to limit the amount of colours to be used or to keep to a colour scheme. So often enamels can be unpleasant because too many colours have been used together without any sense of control or harmony.

Visits to museums nurture ideas for decoration and one can achieve inspiration from natural forms such as plants, insects, crystals and shells. It is a good idea to built up a collection of natural and found objects which attract by their colour, texture or shape, eg, stones, feathers, butterflies, etc. Photography can also be a source of information, especially photomicrographs, eg, photographs taken directly through the microscope by attaching the camera to the lens of the microscope.

One of the simplest ways to design for transparent enamel and foil, is to use coloured cellophane paper and gold and silver metallic paper. The gold or silver can either serve as a complete background or may be cut into sections or shapes and glued on to a paper or card background. Superimpose shapes of the coloured cellophane over the silver and gold. One could build up a warm composition using pink, orange, red and yellow over silver and gold, or a cool one using blues, greens and purple over silver. Use a pentel pen or indian ink with brush or stick to design freely on top of the coloured paper. When the design has been satisfactorily completed it is comparatively easy to translate into enamel using silver and gold foils and transparent enamels.

117 Detail of panel using firescale and glass found on the sea shore

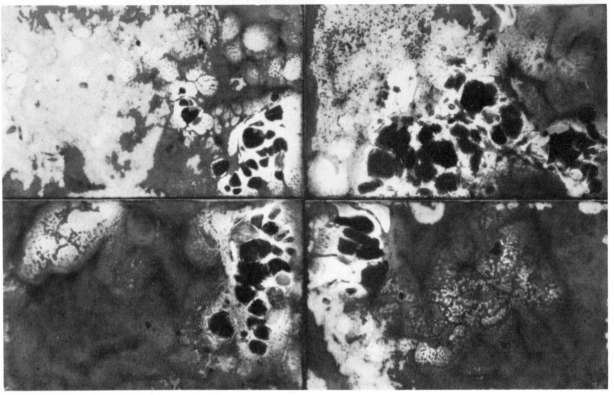

118 Copper panel. Firescale, transparent flux, and glass. *Under the microscope*

119 Sgraffito on copper. *Florentine Night*

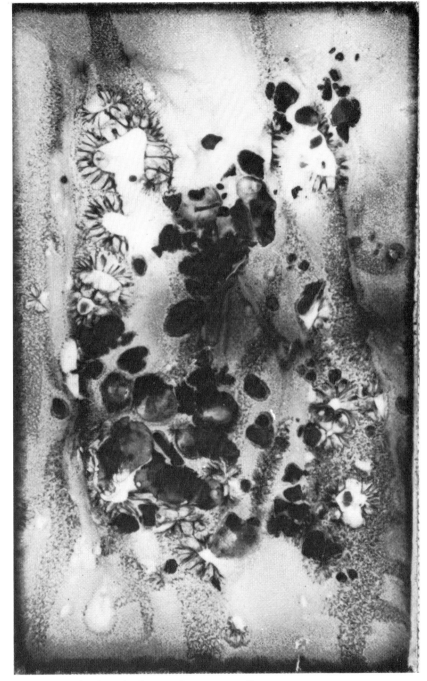

120 *Seaweed*. High fired enamel on copper. Courtesy the Whibley Gallery

121 Kings and Queens

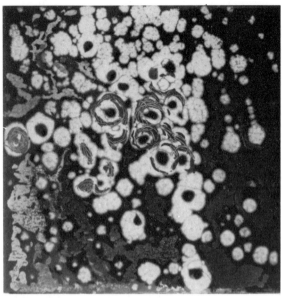

122 Steel panel. Splatter enamel with gold lustre

123 Steel panel. Splatter slush enamel

124 Steel panel. Slush (painted) enamel

125 Steel panel. Slush (painted) enamel

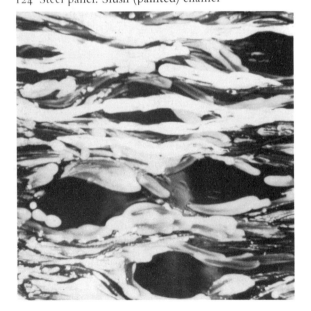
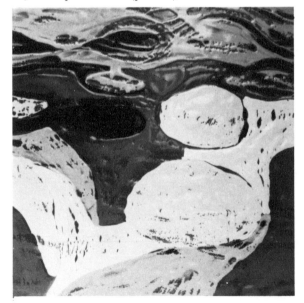

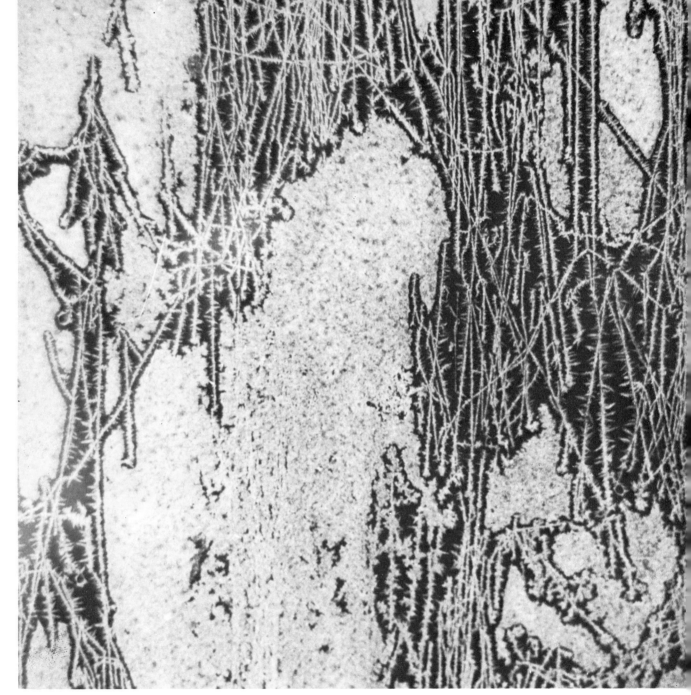

126 Frost patterns on glass

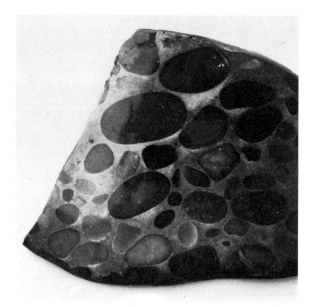

127 Pebbles

128 Polished rock surface

129 Shell (spider conch, Indian Ocean)

130 Shell (textile cone, Indo-Pacific)

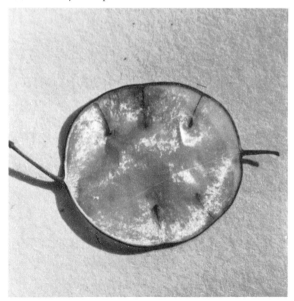

131 Clematis
133 Teazle

132 Globe artichoke flower
134 Honesty seed pod

Suppliers

Great Britain

Enamels

W. G. Ball Limited, Longton Mill, Anchor Road, Longton, Stoke-on-Trent

Blythe Colours Limited, Cresswell, Stoke-on-Trent

Ferro Limited, Wombourne, Wolverhampton

W. J. Hutton (Enamels) Limited, 285 Icknield Street, Birmingham 18

Wengers Limited, Etruria, Stoke-on-Trent

Kilns and enamelling requisites

Art and Crafts Unlimited, 49 Shelton Street, London WC2

W. G. Ball Limited, Longton Mill, Anchor Road, Longton, Stoke-on-Trent

Enamelaire, 61B High Street, Watford, Herts

Wengers Limited, Etruria, Stoke-on-Trent

Enamelling kits

E. J. Arnold and Sons Limited, Butterley Street, Leeds 10

Enamelaire, 61B High Street, Watford, Herts.

Silk Screen equipment

Margros (Marvin Medium), Monument House, Monument Way West, Woking, Surrey

Selectasine Silk Screens Limited, 22 Bulstrode Street, London W1

Steel panels (large)

Vitreous Enamel (Slough), Venture Works, Mill Street, Slough, Middlesex

Copper shapes

H. W. Landon and Brother, 9–12 Bartholomew Road, Birmingham 5

Copper sheet

J. Smith and Sons (Clerkenwell) Limited, 42–54 St John's Square, London EC1

A. Boucher (Metals) Limited, Shelford Place, London N16

Henry Righton and Company Limited, 70 Pentonville Road, London N1

H. Rollet and Company Limited, 6 Chesham Place, London SW1

Foil

Art and Crafts Unlimited, 49 Shelton Street, London WC2

U.S.A.

Enamels

Art-Brite Color and Chemicals, Manufacturers, 19 La Grange Street, Brooklyn, NY 11206

Nesbert L. Cochran, 2540 South Fletcher Avenue, Fernandiha Beach, Florida

Copper sheet

Chase Brass and Copper Company (offices in most cities)

Kilns and enamelling requisites

Electric Hotpak Company, Coltman Avenue at Melrose Street, Philadelphia, Pennsylvania

Ferro Corporation, 4150 East 56th Street, Cleveland, Ohio

Tools and equipment

Allcraft Tool and Supply Company, 11 East 48th Street, New York, NY 10017

Bergen Arts and Crafts, Salem, Massachusetts.

Bibliography

Enamelling
Enameling Principles and Practice, Kenneth F. Bates, World Publishing Company, Cleveland, Ohio

The Enamelist, Kenneth F. Bates, World Publishing Company, Cleveland, Ohio

The Technique of Enamelling, Geoffrey Clarke, Francis and Ida Feher, Batsford, London; Reinhold, New York

Make Your Own Enamels, Jutta Lammer, Batsford, London; Reinhold, New York

The Craft of Enamelling, K. Neville, Mills and Boon

Enameling on Metal, Oppi Untracht, Pitman, London; Chilton Company, Philadelphia

Creative Enamelling and Jewelry-Making, Katharina Zechlin, Oaktree Press

Jewellery
Making Jewellery, K. J. Hartwell, Hulton

Jewellery and Enamelling, Greta Pack, Van Nostrand, New York

Introducing Jewelry Making, John Crawford, Batsford, London; Watson-Guptill, New York

Metalwork
Metalwork and its Decoration by Etching, Oscar Almeida, Mills and Boon

Metalwork and Enamelling, Herbert Maryon, Chapman and Hall

Mobiles
Calder Mobiles and Stabiles, Collins, UNESCO

Making Mobiles, Anne and Christopher Moorey, Studio Vista